SEAGRAVE FIRE APPARATUS
1959-2004 PHOTO ARCHIVE

Kent D. Parrish

Iconografix
Photo Archive Series

Iconografix
PO Box 446
Hudson, Wisconsin 54016 USA

© 2004 Kent Parrish

All rights reserved. No part of this work may be reproduced or used in any form by any means... graphic, electronic, or mechanical, including photocopying, recording, taping, or any other information storage and retrieval system... without written permission of the publisher.

The information in this book is true and complete to the best of our knowledge. All recommendations are made without any guarantee on the part of the author or Publisher, who also disclaim any liability incurred in connection with the use of this data or specific details.

We acknowledge that certain words, such as model names and designations, mentioned herein are the property of the trademark holder. We use them for purposes of identification only. This is not an official publication.

Iconografix books are offered at a discount when sold in quantity for promotional use. Businesses or organizations seeking details should write to the Marketing Department, Iconografix, at the above address.

Library of Congress Control Number: 2004102903

ISBN 1-58388-132-8

04 05 06 07 08 09 6 5 4 3 2 1

Printed in China

Cover and book design by Dan Perry

Copyediting by Suzie Helberg

COVER PHOTO: MODEL TR-74-DA, SERIAL NUMBER 85067 – The Meanstick is currently one of Seagrave's most popular products. This compact quint displays many of the trends in modern fire apparatus design. The climate-controlled cab has seating for six and storage for medical equipment. A heavy fire attack can be accomplished through the 2000-gpm pump. The single rear axle accommodates a 500-gallon water tank and 20-gallon foam cell. There is compartment space for a vast amount of equipment and a full complement of ground ladders. Note the front "brow" scene light. The 75-foot ladder has a 500-pound tip load and the pre-piped waterway is equipped with a remote controlled nozzle. North Oldham, Kentucky was the recipient of this 2002 model with Detroit Diesel Series 60 engine. *Photo by Kent Parrish*

BOOK PROPOSALS

Iconografix is a publishing company specializing in books for transportation enthusiasts. We publish in a number of different areas, including Automobiles, Auto Racing, Buses, Construction Equipment, Emergency Equipment, Farming Equipment, Railroads & Trucks. The Iconografix imprint is constantly growing and expanding into new subject areas.

Authors, editors, and knowledgeable enthusiasts in the field of transportation history are invited to contact the Editorial Department at Iconografix, Inc., PO Box 446, Hudson, WI 54016.

DEDICATION

This book is dedicated to my young son, Logan. While the words "fire truck" only come out as "fa-fa" right now, his admiration of fire apparatus is already evident. May he one day share in the same passion or develop one to call his own.

ACKNOWLEDGEMENTS

This book would not have been possible without the valuable assistance of fellow firefighters and buffs. Their knowledge, personal records, and photographs allowed completion of this volume. I extend special thanks to each individual who contributed to the idea and development of this project. Proper credit is given with each photograph. I would also like to express my appreciation to the firefighters who take the time to proudly pose rigs for photographs. Not only do they appease buffs across the country, but they are also contributing to the historical documentation of fire apparatus. Every effort has been made to accurately chronicle dates, model numbers, and serial numbers in this book. The author welcomes any corrections or clarifications.

The front cover of a vintage Seagrave brochure proudly displays the new Seagrave K-Model. The modern cab-forward was quite an improvement over the early apparatus shown at the bottom. It's hard to imagine that the evolution between these two rigs only spanned 48 years. *Courtesy Dennis Downes collection*

INTRODUCTION

Founded in 1881, Seagrave is the oldest continuous manufacturer of motor fire apparatus in the United States. The history of Seagrave and its classic models, including the famous 70th Anniversary Series, has been much documented and enjoyed. However, little has been done to show the evolution of the modern Seagrave.

Seagrave's "modern era" began in 1959 with the introduction of the cab-forward K-Model. Longer in length than the conventional 70th Anniversary Series, but shorter in wheelbase, advantages included better visibility and improved maneuverability. Made from Seagrave's innovative tube steel frame, the cab measured 72 inches at the front of the door. It would not take long for the K-Models to dominate the production floor at the Columbus, Ohio plant. In 1961, Seagrave unveiled its first elevating platform, the "Eagle." It was available in 65- and 90-foot models. The first unit was produced on a conventional chassis, but a handful of cab-forwards would be delivered.

In 1963, the Four Wheel Drive Auto Company (FWD) acquired Seagrave. During the next two years, production was transferred to Clintonville, Wisconsin and the first diesel powered apparatus rolled out of the plant. The Seagrave Rear Admiral was introduced in 1967. Pilot models were built on FWD chassis with "Cincinnati" cabs and rear steering. The straight frame rear-mount ladders were available in 85- and 100-foot models. Production would soon begin on the low profile 80-inch-wide S-Model cab. After a nice run, the K-Model was replaced by the P-Model in 1969. It was merely an improvement over the original cab-forward. The cab height was lowered by three inches and width was increased to 80 inches.

Seagrave introduced its second elevating platform in 1972, the 85-foot articulating "Astro-Tower." It was available on an extended-wheelbase chassis with single rear axle. Production of the W-Model began five years later. The low-profile cab measured a full 92-inches wide. The front of the cab featured removable access panels outlined with rubber zip gaskets. These allowed for ease of maintenance and would become a symbol of future Seagrave models. In 1978, Seagrave began production of the M-Model. It was a no frills economy pumper originally designed to compete for military contracts. It would later be released to civilian departments as the "Invader." The H-Model, or "Commander," was introduced in the following year. Basically a full height W-Model, it featured jogging window lines and rectangular air intakes on the cab sides. It would become Seagrave's prominent cab until 1988.

During the mid 1980s, Seagrave offered a short-wheelbased 75-foot Rear Admiral quint. Designed as a pumper-based unit, the standard aerial turntable had been deleted and controls were placed at the pump panel. Two H-style stabilizers were at the rear with a single jack underneath the front bumper. In 1987, Seagrave introduced the RA110 heavy-duty ladder. With a 550-pound tip load, the 110-foot rear-mount aerial required the use of four H-style stabilizers. Heavy-duty tractor-drawn aerials were available as well.

In 1988, Seagrave unveiled the J-Model or "Commander II." It displayed a more squared modern look with a cab width of 96 inches, which is now the standard. Present were the familiar front access panels and new vertical air intakes on the cab sides. Improvements were made to the interior as well. By 1990, the L-Model had replaced the W-Model as Seagrave's low-profile offering. Featuring an offset roofline, this cab would be the first offered in only a four-door format. This series has been used exclusively for aerial apparatus.

Seagrave had a very busy year in 1991. Voices from the fire service had been demanding more open space in the cab. The biggest benefit would be ease of communication and a less cramped feel. Seagrave finally unveiled the T-Model or "Marauder." This series would see the most extensive line of options to date. Variations included full-tilt and split-tilt. Both were available with raised roofs and the split-tilt was available with an extended cab allowing seating for up to 10 personnel. A two-door tilt-cab was also available on which other manufacturers could complete specialty apparatus.

In the same year, new NFPA safety standards required two big changes in fire apparatus. One was that all new fire apparatus must have fully enclosed cabs and all personnel had to be in a seated position. Second was that aerial ladders must have a minimum 250-pound tip load rating. The Seagrave Rear Admiral did not meet this requirement and was replaced by the "Patriot." Available were 75- and 100-foot models in rear-mount and tractor-drawn configurations.

Seagrave also unveiled its first tower ladder, the 105-foot "Apollo," in 1991. This heavy-duty unit had an impressive tip load of 1000-pounds dry and 550-pounds while flowing water at any elevation. It was available on either the L- or J-Model chassis. In 1992, Mack ceased production of custom fire apparatus chassis and the popular Baker Aerialscope found itself without a chassis manufacturer. Seagrave would provide the heavy-duty J-Model chassis and fixed cab for the Aerialscope.

In 1996, the "P-75" was introduced. It was an updated pumper-style unit with a 75-foot rear-mount aerial ladder. A year later, the "Meanstick" was unveiled. This has proved to be one of Seagrave's most popular modern designs. It is a true quint with a 75-foot rear-mount aerial, all on a short-wheelbase chassis with single rear axle. In 1999, the "Force," a more compact heavy-duty aerial ladder, was introduced. A year later, the "Flame" was unveiled. It was advertised as an economy pumper, targeted at smaller communities with tight budgets who wanted a custom rig. In 2001, Seagrave announced the "Specialist," its own line of custom heavy rescue vehicles.

It is important to note that FWD and Seagrave are synonymous. Generally, FWD nameplates can be found on a custom Seagrave in three instances: a chassis supplied to another manufacturer for completion of a specialty apparatus, a custom Seagrave on a four-wheel or six-wheel-drive chassis or a custom Seagrave chassis supplied for the Baker Aerialscope. Today, the FWD nameplate has all but disappeared from new Seagrave apparatus. Four-wheel and six-wheel drive are no longer options, Seagrave is now producing its own heavy rescue vehicle, and the Aerialscope line has been purchased by Seagrave.

In order to fully explore the progression of the modern Seagrave, this volume will only focus on their custom apparatus. However, it is important to note that Seagrave did produce apparatus on commercial chassis until 1990. Seagrave also licensed its name to three other companies in its Modern Era. King-Seagrave of Canada, Timmons Metal Products Company (Timpco) of Columbus, Ohio, and Clintonville Fire Apparatus (CFA) produced apparatus under the Seagrave name. The vast majority of these units were built on commercial chassis, but King-Seagrave and CFA did produce a handful of custom rigs. Seagrave also produced apparatus on FWD and International custom chassis with "Cincinnati" cabs.

At a time when many of the great names in fire apparatus have disappeared, Seagrave shows no signs of slowing down. In 2003, James Hebe and a group of private investors, managed through Ballamor Capital Management, Inc., acquired the company. A new Canadian partner was found in Altamonte. Apparatus on commercial chassis is once again being marketed. Aerialscope production was moved from Richmond, Virginia to the Clintonville plant. It was announced that the venerable Detroit Diesel engines would be phased out—Caterpillar motors would become standard and Cummins would be available as an option. Also, the J-Model chassis was completely redesigned.

In 2004, Seagrave announced an entirely new product line. Two new custom chassis were introduced, the "Concorde" and the "Attacker." The "Tower Max," a new aerial tower with 85 and 104-foot models, was debuted. Seagrave also acquired Classic Fire, a Florida company who produced smaller apparatus on commercial chassis. Two exclusive options were developed, a new air-ride front suspension and Seagrave's own fire pump, the "Triton." With time proven models, a new product line, aggressive marketing, and construction of an entirely new plant in Clintonville, Seagrave will continue to stake its claim as the "Greatest Name in Fire Apparatus."

Seagrave has utilized three formats of model numbers throughout the Modern Era, detailed in the charts here. For example, during the first period, a "900-KA-1000-85/3" was a cab-forward quint with 1000-gpm pump, 85-foot three-section aerial, and Seagrave 906 gasoline engine. During the second period, a "PB-25068" was a P-Model 4x2 pumper with 1500-gpm pump, no aerial, and Detroit Diesel 8V-71N engine. In the third and present format, a "TP-55-DA" is a tilt-cab tandem axle job with 1500-gpm pump, 100-foot rear-mount ladder with 500-pound tip load, and Detroit Diesel Series 60 engine.

1959 to 1969

Engine		Cab	Chassis		Pump		Aerial	
531	Seagrave 531	K	A		Aerial	500	65-foot/3-section
550	Waukesha 140-GZ		B		Pumper	750	75-foot/3-section
590	Hall Scott 590		AB		Quint	1000	85-foot/3-section
800	Waukesha F-817-G		BH		Heavy chassis	1250	100-foot/3-section
900	Seagrave 906		E		Eagle platform	1500	100-foot/4-section
935	Hall Scott 6156-FE		EP		Pitman Snorkel		
350D	Detroit Diesel 8-V71		EB		Eagle Quint		
702	GMC 702 Twin Six V12		EPB		Snorkel quint		
			Q		Quad		
			S		Special chassis		
			T		Tractor-drawn		

1969 to 1982

Cab	Chassis		Drive		Pump		Aerial		Detroit Diesels		Cummins	
K	B	Pumper	2	4x2	2	750	0	no aerial	53	6-L-71TAC	77	NTCC300
P	A	Mid-mount Aerial	4	4x4	3	1000	5	85'	54	6V-92TAC	78	NTCC350
S	R	Rear-mount Aerial	5	tandem	4	1250	7	100'	55	8V-92TAC	79	NTC-250
W	T	Tractor-drawn			5	1500	8	100' HD	56	6-71N	85	Pwr Torque 270
H	C	Pumper chassis only			6	1750-sp			57	6-71TA	86	NTC-290
	S	Aerial chassis only			7	1750			62	6V92	88	NTC-350
	X	Tractor only			8	2000			63	8V92	97	NTC-200
					9	PTO			64	6V-92TA	98	VT-225
					0	n/A			65	8V-92TA	99	NHTF-295
									66	6V-71N		
									67	6V-71T, 8V-71TA	9	* Hall Scott 906
									68	8V-71N	29	* Waukesha F-817-G

* Denotes gasoline engine models

1982 to Present

Cab	Chassis		Pump		Aerial		Detroit Diesels		Cummins	
H	B	Pumper, 4x2	3	1000	1	Special	DA	6-71N, Series 60	CA	VT-225
J	C	Chassis only, 4x2	4	1250	2	105' tower	DB	6-71T	CB	VT-225
L	D	Pumper, 4x4	5	1500	4	85' RA, 75' Pat/500lb	DC	8V-71N, Series 40	CC	NTC-250, PT-270
T	E	Pumper-tanker, 6x4	6	1750	5	85' RA, 100' Pat/500lb	DD	8V-71TA, Series 50	CD	NTC-290, NTC-300
	F	Chassis only, 4x4	7	2000	6	100'/4-JK, 100' Pat/250lb	DE	8V-71TAC	CE	NTCC-290, NTCC-300
	G	Chassis only, 6x4	8	2250	7	100' Rear Admiral	DF	6V-92TA	CF	NTC-350
	H	Chassis only, 6x6	9	PTO	8	100' Heavy-Duty	DG	6V-92TAC	CG	NTCC-350
	J	Pumper-tanker, 6x6	0	n/a	9	RA110	DH	8V-92TA	CH	NTC-400
	P	RM aerial, 6x4			0	No aerial	DJ	8V-92TAC	CJ	L10-240
	R	RM aerial, 4x2					DK	67-1TAC	CN	N-14
	T	Tractor-drawn								
	X	Tractor only								

MODEL 531-KB-1000, SERIAL NUMBER L-750 – There has been some debate as to who received the first Seagrave cab-forward design. Factory records show that a unit for Mehlville, Missouri was entered into the logs on December 12, 1958. More than likely, this was the first K-Model ordered. Painted solid white, it had a Seagrave 531 gasoline engine, 1000-gpm pump, foam system, and 500-gallon water tank. *Photo from Dennis J. Maag collection*

MODEL 531-KB-1000, SERIAL NUMBER L-995 – It is believed that this pumper for Jackson, Tennessee was actually the first Seagrave cab-forward completed and delivered. Legend has it that Seagrave wanted to get the "red" rigs out first because they photographed better and made better advertisements. Jackson's rig had a Seagrave 531 gasoline engine and 1000-gpm pump. When photographed, it was sitting proudly on the front apron of the station, ready to respond. *Photo by Richard Adelman*

MODEL 935-KT-100, SERIAL NUMBERS L-2455 & L-2456 – Portland, Oregon must have liked the new Seagrave cab-forward because they received a total of 13 with Hall Scott 6156-FE gasoline engines in 1959, including two of these open-cab 100-foot tractor-drawn aerials. Three-section aerials were considerably long, thus the tiller seats on these rigs were designed to swing over the ladders if aerial operations were necessary. Road sanders were located just forward of the tractor's rear fenders. *Photo from Richard Adelman collection*

MODEL 531-KA-75, SERIAL NUMBER L-2920 – This open-cab 75-foot mid-mount aerial was originally delivered to West View, Pennsylvania in 1960. It was equipped with a Seagrave 531 gasoline engine, 250-gpm booster pump, 250-gallon water tank, and hose reel. Often, a canvas cover would be factory installed for the delivery trip. Some departments retrofitted their open-cabs with canvas tops at a later date. When photographed, this rig was serving its second home in Avis, Pennsylvania. *Photo by Rick Rudisill*

MODEL 800-KB-1000, SERIAL NUMBER L-7175 – This 1000-gpm pumper was delivered to Westlake, Ohio in 1960. It was powered by a Waukesha F-817-G gasoline engine. Although common today, red apparatus with white upper cabs were somewhat flashy at this time. This view nicely displays the cab-forward's canopy cab configuration. Note the small equipment box at the pump panel. *Photo from Richard Adelman collection*

11

MODEL 531-KB-1000, SERIAL NUMBER L-6425 – The Joliet Army Ammunition Plant in Illinois took delivery of this pumper in 1960. It had a Seagrave 531 gasoline engine, 1000-gpm pump, and 300-gallon water tank. Advantages of the cab-forward over the conventional models included better visibility and maneuverability. By sitting higher up and ahead of the front wheels, the driver had full view of the road. *Photo by Garry Kadzielawski*

MODEL 900-KT-100, SERIAL NUMBER M-8400 – This 100-foot tractor-drawn aerial with Seagrave 906 gasoline engine was delivered to the Excelsior Hook & Ladder Company No. 1 of Freeport, New York in 1961. It was probably the most elaborate and ornate cab-forward produced to date. Included were extra warning equipment, chrome wheels and accents, and plenty of gold leaf. The fixed tiller position had a unique rollback cab, which could be kept fully enclosed or semi-open. *Photo from Richard Adelman collection*

MODEL 800-KB-1000, SERIAL NUMBER M-3375 – Lima, Ohio received this solid white 1000-gpm pumper with Waukesha F-817-G gasoline engine in 1961. Note the addition of narrow hand tool compartments and all-weather tires. The phrase "purchased with your income tax" adorned the cab doors, giving thanks to the community for its support. *Photo from Richard Adelman collection*

MODEL 900-KA-100, SERIAL NUMBER N-1690 – The Hamilton Fire Company in Hamilton Township, New Jersey received this striking 1962 100-foot mid-mount aerial in 1962. The rig was painted in an olive green motif. The red wheels were adorned with chrome accessories. Gold leaf lettering and striping had red accents. The apparatus was powered by a Seagrave 906 gasoline engine. When this photo was taken, the unit had been equipped with a more modern light bar and a Stokes basket. *Photo by Scott Mattson*

MODEL 531-KB-750, SERIAL NUMBER N-8365 – The Washington Fire Company in Delanco, New Jersey received this beautiful cream-colored 1963 open-cab pumper. It had a Seagrave 531 gasoline engine, 750-gpm pump, and 350-gallon water tank. A bell was mounted in a bracket at the center of the front bumper and the seating featured red upholstery. Note the handy placement of the pick head axe. *Photo by Scott Mattson*

MODEL 800-KB-750, SERIAL NUMBER N-9100 – This sharp looking 1963 K-Model went to the Relief Fire Company in Burlington Township, New Jersey. Painted white and red, the apparatus had a Waukesha F-817-G gasoline engine, 750-gpm pump, and 300-gallon water tank. A Bangor ladder was carried in a rack mounted over the hose bed. Note the small equipment box at the pump panel. *Photo by Mike Martinelli*

MODEL 702-KB-1000, SERIAL NUMBER P-4865 – The two rigs on this page represent engine distinction and milestone. This 1964 delivery to Milwaukee, Wisconsin was the only custom Seagrave cab-forward to be powered by a GMC 702 Twin Six V-12 gasoline engine. Engine 49 was equipped with a 1000-gpm pump and carried 300 gallons of water. Note the cutout for soft suction hose in the lower front center of the cab. *Photo by Bill Friedrich*

MODEL 350D-KB-1250, SERIAL NUMBER P-5250 – By now, production was being moved from Columbus, Ohio to Clintonville, Wisconsin as a result of the FWD acquisition. The first diesel powered Seagrave to roll out was this 1964 K-Model with Detroit Diesel 8V-71N motor for the Community Fire Protection District in Missouri. Diesels could be identified by the bulge in the lower engine compartment. The apparatus was equipped with a 1250-gpm pump and 300-gallon water tank. *Photo by Dennis J. Maag*

MODEL 900-KT-100, SERIAL NUMBER P-9805 – Volunteer departments on Long Island, New York had long been known for ornate apparatus. This 1965 open-cab 100-foot tractor-drawn aerial for Mamaroneck was no exception. Powered by a Seagrave 906 gasoline engine, the apparatus had a four-section aerial with fixed tiller position. The trailer was equipped with special roll-up compartments directly underneath the aerial. Note the extra warning beacon positioned at the middle of the trailer. *Photo by John Toomey*

MODEL 800-KEPB-1000-65, SERIAL NUMBER P-6200 – Grosse Pointe, Michigan received this impressive K-Model in 1965. While most Seagrave Snorkels had bodywork by another manufacturer, this rig was built completely by Seagrave. The apparatus had a Waukesha F-817-G gasoline engine, 1000-gpm pump, 250-gallon water tank, and 65-foot Pitman boom. Ground ladders were carried in enclosed boxes on both sides. Originally painted red, the unit was repainted lime. *Photo by Bill Friedrich*

MODEL 800-KB-1000, SERIAL NUMBER P-5500 – Detroit, Michigan operated an entire fleet of Seagrave Anniversary Series rigs at one time. Although the department continued to receive the conventional models until 1968, this K-Model was assigned to Engine 5 in 1965. It had a Waukesha F-817-G gasoline engine, 1000-gpm pump, and 300-gallon water tank. Note the canopy cab extension—just a rear wall and two extra doors away from a fully enclosed model. Forward facing flip seats in the jump area could carry additional firefighters. *Photo by Chuck Madderom*

MODEL 800-KA-100/4, SERIAL NUMBER Q-4180 – This 100-foot mid-mount aerial was delivered to Denver, Colorado in 1966. It was powered by a Waukesha F-817-G gasoline engine and had hydraulic stabilizers. The traditional Denver color scheme was white body with red wheels. A tarp covered the aerial to protect it from the harsh weather conditions typical of the "Mile High City." When photographed, this rig was serving as a reserve unit. *Photo by Dennis J. Maag*

MODEL 900-KEB-1000-90, SERIAL NUMBER P-5830 – The first Seagrave "Eagle" elevating platform was produced on a 70th Anniversary Series chassis. It was the only 70-foot conventional model produced. Melrose Park, Illinois took delivery of this K-Model with 90-foot mid-mount boom in 1966. The Young Spring and Wire Company built the three-section articulating platform exclusively for Seagrave. Truly a beast in its time, the Melrose Park rig had a Seagrave 906 gasoline engine and a 1000-gpm pump. Just five K-Model 90-foot Eagles were produced. *Photo by Bill Friedrich*

MODEL 900-KB-1000, SERIAL NUMBER Q-2795 – The Manassas Volunteer Fire Company in Virginia received this open-cab pumper in 1966. Powered by a Seagrave 906 gasoline engine, the unit was equipped with a 1000-gpm pump and carried 500 gallons of water. The body was painted red with white upper cab and wheels. A partition at the lower pump panel contained a section of pre-connected hose. Note the electronic siren speaker and strobe lighting on the front bumper. *Photo by Mike Sanders*

MODEL 800-KT-85/4, SERIAL NUMBER Q-4323 – The small town of Athens, Ohio received a pair of 1967 K-Models, a pumper and the last 85-foot tractor-drawn aerial built by Seagrave. It had a four-section ladder, making for an extremely maneuverable rig. The fixed tiller position was left open. Ladder 1 was powered by a Waukesha F-817-G gasoline engine. *Photo by Steve Hagy*

MODEL 800-KA-65, SERIAL NUMBER Q-4246 – Seagrave produced very few cab-forward 65-foot mid-mount aerials. Wichita, Kansas received this 1967 model powered by a Waukesha F-817-G gasoline engine. The department had also received a similar unit a few years earlier. The unusual rear design gave the rig a short appearance. A small offset compartment was placed just back of the rear fender rather than a conventional cabinet. *Photo by Bill Friedrich*

MODEL 800-KB-1000, SERIAL NUMBER Q-4192 – This 1967 crash unit was delivered to the Orlando International Airport in Florida. Powered by a Waukesha F-817-G gasoline engine, it could pump 1000 gallons-per-minute through its main pump and the big turret mounted on the cab roof. The apparatus carried 500 gallons of water and 250 gallons of "light water" or foam. Unusual box-type compartments were affixed to the driver's side of the body. *Photo by Garry Kadzielawski*

MODEL FF9-4368, SERIAL NUMBER R-4341 – Seagrave introduced the popular Rear Admiral line of aerials in 1967, available in 85- and 100-foot models. The first units were built on FWD chassis with "Cincinnati" cabs and rear steering. "Rear-steers" could be identified by the single rear wheels with hubs similar to the front wheels. Chicago, Illinois purchased this 1968 demonstrator model. Given a FWD model number, the low-profile cab on this rig would come to be known as the Seagrave S-Model. The unit was assigned to Truck 1 and had a Detroit Diesel 8V-71N engine. *Photo by Chuck Madderom*

MODEL 900-KAR-100/4, SERIAL NUMBER R-4339 – Seagrave delivered this 100-foot Rear Admiral to Bryn Mawr, Pennsylvania in 1968. Although given a K-Series model number, it was actually an open-cab S-Model. It was one of two consecutive open-cab rear-mounts that Seagrave built. When photographed, this rig was serving its second home in Nesquehoning, Pennsylvania. Powered by a Seagrave 906 gasoline engine, it retained the original gold-over-dark green paint job and had been equipped with modern light bars and power cord reels. *Photo by Mike Martinelli*

MODEL 350D-KQ-1000, SERIAL NUMBER R-4452 – Kansas City, Missouri received this interesting quad in 1969. Equipped with a Detroit Diesel 8V-71N engine, 1000-gpm pump, and 300-gallon water tank, the apparatus was laid out so that it could later be retrofitted with a mid-mount aerial ladder. Intentions never came to fruition and the rig remained a quad. *Photo by Chuck Madderom*

MODEL KB-23056, SERIAL NUMBER R-4528 – The Greater Springfield Volunteer Fire Department in Fairfax County, Virginia received this sharp looking open-cab pumper in 1969. The front fenders, rear compartments, and wheels sported a black scheme, while the rest of the apparatus was painted white. Powered by a Detroit Diesel 6-71N engine, the rig had a 1000-gpm pump and carried 500 gallons of water. *Photo by Chuck Madderom*

MODEL KA-20568, SERIAL NUMBER R-4454 – This clean looking 85-foot mid-mount aerial was delivered to Prince William County, Virginia in 1969. It had a Detroit 8V-71N engine and hydraulic outriggers. The body was equipped with numerous transverse compartments underneath the ladder. D-ring compartment door latches would become more prevalent than the old-style paddle handles. Note the classic Pompier ladder carried on the side. *Photo by Mike Sanders*

MODEL PB-20468, SERIAL NUMBER R-4536 – Seagrave introduced the P-Model in 1969. An improvement over the original cab-forward, cab height was lowered by three inches and width was increased to a full 80 inches. Diesel engines would soon become standard. The West Overland Fire Protection District in Missouri received the first unit. It had a Detroit Diesel 8V-71N engine and 1250-gpm pump. The warning beacon had been lowered in front of the cab. *Photo from Richard Adelman collection*

MODEL PT-20756, SERIAL NUMBER R-4628 – The first P-Model tractor-drawn aerial was this 100-foot model delivered to Anne Arundel County, Maryland in 1970. Powered by a Detroit Diesel 6-71N engine, the rig featured a pale yellow and white color scheme with black wheels. The trailer was equipped with a fully enclosed tiller cab and hydraulic outriggers. *Photo by Howard Meile*

MODEL PB-23066, SERIAL NUMBER R-4757 – The East Dover Volunteer Fire Company in Toms River, New Jersey received this compact rig in 1971, one of only two 55-foot Snorkels completely finished by Seagrave. Snorkel-equipped units usually had bodywork completed by Pierce. Powered by a Detroit Diesel 6V-71N engine, the apparatus had a 1000-gpm pump and 500-gallon water tank. *Photo by John Toomey*

MODEL SB-23029, SERIAL NUMBER R-4535 – The only S-Model pumper built by Seagrave was this 1971 delivery to the Chemical & Salvage Co. No. 2 of Lindenhurst, New York. Powered by a Waukesha F-817-G gasoline engine, the low profile apparatus had a 1250-gpm pump and 500-gallon water tank. Note the mid-mounted shield, which deflected wind up over the exposed hose bed. Typical of East Coast departments was the sight of turnout gear hanging from the sides. *Photo by John Toomey*

MODEL PB-23056, SERIAL NUMBERS R-4630 THROUGH R-4633 – Baltimore, Maryland received this group of P-Model pumpers in 1971. They were among the earliest modern Seagraves to have four-door fully enclosed cabs. The spacious cabs could seat a crew of seven. Powered by Detroit Diesel 6-71N engines, the pumpers were equipped with 1000-gpm pumps and 300-gallon water tanks. Baltimore had switched its color scheme from white and red to white and orange. *Photo by Howard Meile*

MODEL SS-20068, SERIAL NUMBER R-4537 – Seagrave introduced its second generation of modern aerial platforms in 1972. The "Astro-Tower" was basically a Rear Admiral, but with a longer wheelbase and a 70-foot articulating platform instead of the ladder. The Trump Company of Canada produced the boom for Seagrave. Greenville, South Carolina received this delivery. It was built on the S-Model low-profile chassis and had a Detroit Diesel 8V-71N engine. *Photo by Dave Organ*

MODEL PQ-23568, SERIAL NUMBER C-73214 – The Walnut Street Fire Company in Madison, Indiana received this 85-foot mid-mount quint in 1972. Company 4 was powered by a Detroit Diesel 8V-71N engine. Painted shamrock green, the unit was equipped with a 1250-gpm pump and carried 200 gallons of water. A booster reel adorned the driver's side of the rig. By now, hydraulic outriggers were standard. Note the gear hanging from the rear. This was a trait normally seen in East Coast volunteer fire departments. *Photo by Greg Stapleton*

MODEL PB-23029, SERIAL NUMBER B-73132 – The Middletown Township Fire Company in Port Monmouth, New Jersey received this open-cab pumper in 1973. At a time when departments were ordering closed cabs and diesel engines, this rig was specified with an open-cab and a Waukesha F-817-G gasoline engine. A new option was a brushed or polished stainless steel pump panel. *Photo by Scott Mattson*

MODEL SR-20768, SERIAL NUMBER A-73010 – This 100-foot Rear Admiral with Detroit Diesel 8V-71N engine was delivered to Green Haven, Maryland in 1973. Maryland apparatus have long been noted for striking color schemes and lots of accessories. This rig was a combination of white and forest green. A plethora of portable scene lights and nozzles was mounted to the upper cab. Truck Company 14 could not be missed on the roadway with two roof beacons, three Mars lights, eight red flashers, and a Federal Q2B siren up front. *Photo by Chuck Madderom*

MODEL PT-20768, SERIAL NUMBER C-73199 & MODEL PB-23068, SERIAL NUMBER D-73317 – Cincinnati, Ohio had been a longtime Seagrave customer. In an unfortunate sign of the times, the city began to encounter frequent civil disobedience. The fire department put out specifications for special "riot-proof" apparatus. The fully enclosed P-Model 100-foot tractor-drawn aerials and pumpers were designed to protect firefighters and to enclose as much loose equipment as possible during civil unrest. Truck 16 received this tractor-drawn aerial in 1973 and Engine 2 operated this 1974 1000-gpm pumper with 300-gallon water tank. Both were powered by Detroit Diesel 8V-71N engines. *Photos by Steve Hagy*

MODEL SR-20799, SERIAL NUMBERS D-75026 THROUGH D-75043 – New York City received 18 1974 100-foot Rear Admirals with fully enclosed four-door cabs and Cummins NHTF-295 engines. They were unique in that they were equipped with an additional compartment on the officer's side where a seventh crewmember could ride. A deck gun rode atop the extra cab. Note the vertical exhaust stack directly behind the cab. *Photo by John A. Calderone/Fire Apparatus Journal*

43

MODEL PC-20068, SERIAL NUMBER E-74089 – Occasionally, Seagrave would supply a custom chassis on which another manufacturer would complete a specialty apparatus. Islip, New York specified this air and light unit on a 1975 P-Model chassis with Detroit Diesel 8V-71N engine. Welch performed the bodywork. The unit was equipped with an air cascade system, large generator, and enough candlepower to light up a football field. *Photo by Mike Martinelli*

MODEL PA-20766, SERIAL NUMBER E-75060 – Santa Clara, California received this white-over-lime 100-foot mid-mount aerial in 1975. Powered by a Detroit Diesel 6V-71N engine, the apparatus was equipped with extra compartmentation. Yellow and red beacons adorned the cab roof along with dual siren loudspeakers on the front bumper. Seagrave would produce just a handful more mid-mount aerials before phasing them out. *Photo by Steve Hagy*

MODEL SS-55068, SERIAL NUMBER F-74116 – The S-Model low-profile chassis was an ideal choice for aerial platforms. Dozens of Snorkel and Grove/LTI products found their way onto Seagrave chassis. Clayton, Ohio operated this 1975 model with an 85-foot LTI tower ladder and Detroit Diesel 8V-71N engine. Large cylinders provided breathing air to the bucket. Warning equipment was later upgraded on this apparatus and NFPA reflective striping was added. *Photo by Greg Stapleton*

MODEL PT-20756, SERIAL NUMBERS E-76060 THROUGH E-76066 – This group of P-Model 100-foot tractor-drawn aerials went to Boston, Massachusetts in 1976. They were powered by Detroit Diesel 6-71N engines. The fully enclosed four-door cabs could seat up to seven personnel and the unusual front bumpers were heavily reinforced. Two tall warning beacons and dual siren speakers graced the cab roofs. A ventilation stack extended from the engine compartment through the cab roof. Note the bubble window in the tiller cab. *Photo by Bill Friedrich*

MODEL PB-44068, SERIAL NUMBER F-73518 – Kings Park, New York favored four-wheel drive apparatus. This 1976 delivery was no exception. Engine 41-4 had a Detroit Diesel 8V-71N engine, 1250-gpm pump, and 750-gallon water tank. FWD nameplates graced the front of the cab. A new trend was the use of pre-connected "cross-lay" attack lines over the pump panel, which played out nicely to either side of the apparatus. Note the forward facing flip-down "jump" seat. *Photo by Mike Martinelli*

MODEL WB-25068, SERIAL NUMBER H-73708 – Seagrave introduced the W-Model in 1977. The low-profile cab featured a full 92-inch width and extra large windows. The face of the cab featured removable access panels outlined with rubber zip gaskets. The front defines Seagrave apparatus to this date. Detroit, Michigan received the first delivery. Powered by a Detroit Diesel 8V-71N engine, the compact rig had a 1500-gpm pump and carried 500-gallons of water. Note the Seagrave script just under the driver's side window. Detroit was among many departments to switch to lime colored apparatus for higher visibility while responding to emergencies. *Photo by Chuck Madderom*

MODEL PT-20788, SERIAL NUMBERS G-76083 THROUGH G-76093 – Los Angeles City received this batch of 100-foot tractor-drawn aerials in 1977. Powered by Cummins NTC-350 diesel engines, the apparatus had open-cabs and tiller seats. Amber flashers graced the front of the cabs along with a Federal Q2B siren and single air horn. The trailers were loaded with ventilation fans and wooden ladders, which were still preferred by West Coast departments. *Photo by Garry Kadzielawski*

MODEL MB-23098, SERIAL NUMBER H-95224 – Seagrave began to produce the M-Series pumpers in 1978. This was a line of no frills economy apparatus originally designed to compete for military contracts. Constructed of square sheet metal, the cab had a very basic appearance and a plain interior. Powered by a Cummins VT-225 engine, this 1978 model was equipped with a 1000-gpm pump and 500-gallon water tank. Painted "school bus yellow," it was assigned to the United States Naval Shipyard in San Diego, California. *Photo by Garry Kadzielawski*

MODEL SR-20768, SERIAL NUMBER H-75211 – This pretty 100-foot Rear Admiral with pre-piped waterway was delivered to the Liberty Fire Company in Middletown, Pennsylvania in 1978. The apparatus was powered by a Detroit Diesel 8V-71N engine and featured a baby blue paint scheme with white upper cab and accents. The jump seat area was equipped with forward facing flip seats. Note the cord reel and large scene lights affixed to the upper body. *Photo by John Toomey*

MODEL PT-20756, SERIAL NUMBERS J-76110 THROUGH J-76114 – New York City continued to place large orders with Seagrave. The FDNY received five of these 1978 100-foot tractor-drawn aerials with fully enclosed four-door cabs and Detroit Diesel 6-71N engines. The tiller cabs were enclosed as well. Note the screen placed over the middle cab window. These were installed to protect firefighters in the jump seats during civil unrest and while in "rough" parts of town. *Photo by John A. Calderone/Fire Apparatus Journal*

MODEL PB-25068, SERIAL NUMBER H-75752 – Malverne, New York received this pumper with four-door fully enclosed cab in 1979. Powered by a Detroit Diesel 8V-71N engine, the apparatus was painted chrome yellow and was equipped with a 1500-gpm pump and 500-gallon water tank. It has since been repainted white-over-red. Note the busy pump panel and the big Stang gun aimed over the top of the cab. *Photo by John Toomey*

MODEL WT-20768, SERIAL NUMBERS H-76102 & J-76115 – San Francisco, California received the only W-Model tractor-drawn aerials built by Seagrave. Delivered in 1978 and 1979, the 100-foot models were powered by Detroit Diesel 8V-71N engines. The tiller cabs were of unusual design. Equipped with no doors, the semi-enclosed cabs angled towards the rear of the apparatus with a V-shaped entry. The tillerman had to lift his legs up and slide into the cab. A canvas cover kept the rain out. The trailers were loaded with wooden ground ladders. *Photo by Bill Friedrich*

MODEL HT-20768, SERIAL NUMBER J-76125 – Seagrave started production of the H-Model or "Commander" in late 1978. It would be Seagrave's prominent cab throughout the next decade. The cabs featured jogging window lines and rectangular air intakes. Fairfax County, Virginia took delivery of this 1979 100-foot tractor-drawn aerial with Detroit Diesel 8V-71N engine. The enclosed tiller cab was equipped with an exterior "buddy" seat. This is where an instructor would sit while training a new tillerman or vice versa. *Photo by Mike Sanders*

MODEL WB-24068, SERIAL NUMBER K-73991 – Chicago Heights, Illinois received this W-Model pumper in 1980. Engine 653 was equipped with a 1250-gpm pump, 500-gallon water tank, and Detroit Diesel 8V-71N engine. Of course, it was painted "Chicago-style" black-over-red. The low profile rig was built on the long pumper chassis and had a very lean appearance. *Photo by Garry Kadzielawski*

MODEL PC-20068, SERIAL NUMBER G-74134 – Seagrave delivered numerous P-Model chassis to Western States Fire Apparatus, who completed special order units for departments in the Northwest United States. Many departments in this territory preferred front-mount pumps. Hillsboro, Oregon received this 1980 job in which the 1250-gpm pump and panel were integrated into the front of the cab. The rig was equipped with "pump and roll" capabilities with a bumper cage for protection during wildfire operations. The body carried numerous compartments and 1000 gallons of water. *Photo by Shane MacKichan*

MODEL PS-50068, SERIAL NUMBER K-74157 – Another Northwestern department operated this strange hybrid. In what may be a true oddity, it is a custom 1980 Seagrave P-Model chassis on which Sutphen completed the bodywork. Burlington, Washington specified this 1500-gpm tandem axle rig with 2200-gallon water tank. Legend has it that the department would accept no less than the 2200-gallon water tank, which Seagrave was unwilling to build. A compromise was reached and Seagrave provided the chassis on which Sutphen completed the body. Both of the Northwestern jobs on this page had Detroit Diesel 8V-71N engines. *Photo by Shane MacKichan*

MODEL PB-25068, SERIAL NUMBER K-79053 – One of the most impressive foam pumpers ever built by Seagrave was this 1980 1500-gpm delivery with Detroit Diesel 8V-71N engine to the Fairfax Volunteer Fire Department in Virginia. The rig carried 400 gallons of water and 600 gallons of foam. The big turret on top of the roof could flow 1000 gallons-per-minute. Additional forward facing seats were recessed into the pump housing and the rear was equipped with slanted "beavertail" compartments. A trend that would not become common for several more years was the use of pre-connected "bumper lines." *Photo by Chuck Madderom*

MODEL HB-40-DC, SERIAL NUMBER P-79304 – The Leesburg Volunteer Fire Company in Virginia received this big tandem axle pumper-tanker in 1981. It had a Detroit Diesel 8V-71N engine, 1250-gpm pump, and 2000-gallon water tank. Dual Mars lights graced the front along with red flashers recessed into the side of the bumper. This lower zone warning would eventually be required by the NFPA. Note the side-mounted access ladder, which could be utilized to gain access to the tank fill and deck gun platform. *Photo by Mike Sanders*

MODEL HR-20757, SERIAL NUMBER N-75392 – Because of excessive wear and tear and heavy run volumes, FDNY apparatus do not see lengthy front-line service, usually about 10 years. After stints in reserve or spare status, most are eventually sold as surplus. They are then purchased and refurbished by other departments across the country. This 1982 100-foot Rear Admiral with fully enclosed four-door cab was assigned to Ladder 56. *Photo by Ben Greif*

MODEL HB-40-DB, SERIAL NUMBER 77626 – Although the majority of Seagrave business is new apparatus, the "refurb shop" remains quite busy. The Edgewood Fire District in Louisville, Kentucky purchased FDNY Ladder 56, pictured above, once it became surplus. It was sent to Seagrave, where the H-Model cab was refurbished and the Detroit Diesel 6-71TA engine was overhauled. The aerial and body were removed from the chassis. A new 1500-gpm pump, 500-gallon water tank, and quad body were installed. The "new" unit was completed in 1997. *Photo by Kent Parrish*

MODEL HB-40-DC, SERIAL NUMBER P-79304 – This 1982 H-Model pumper with fully enclosed four-door cab went to Hampstead, Maryland. Engine 22 had a Detroit Diesel 8V-71N engine, 1250-gpm pump, and 750-gallon water tank. The rig was painted solid white with red wheels. A Mars light graces the front of the cab along with dual "tear-drop" lights on each corner of the bumper. Note the early Seagrave flame graphic between the driver and crew windows. *Photo by Howard Meile*

MODEL PT-20757, SERIAL NUMBERS N-76161 & N-76162 – Baltimore City, Maryland was still a consistent Seagrave customer. The department received several batches of 100-foot tractor-drawn aerials with enclosed tiller cabs in 1982. Having served Truck 4 and 12, this unit was powered by a Detroit Diesel 6-71TA engine. A large warning light/siren combination graced the cab roof. Although the H-Model was dominating the production floor at Seagrave, many departments were still ordering P-Models. However, production would drop off sharply after this year and the last unit would be delivered in 1985. *Photo by Howard Meile*

MODEL HC-00-DC, SERIAL NUMBER Q-74191 – This big rig was delivered to the Mill Creek Fire Company in Marshalltown, Delaware. The department specified a 1983 Seagrave H-Model chassis on which Ranger constructed a walk-in heavy rescue body. The apparatus was painted a striking white-over-green. It has since been repainted and now serves the Rock Community Fire Protection District in Missouri. *Photo by John Toomey*

MODEL HJ-40-DB, SERIAL NUMBER Q-79324 – The Bowling Green/Warren County Airport in Kentucky operates this one of a kind 1983 H-Model crash truck with Detroit Diesel 6-71T engine. The 1250-gpm pump has "pump and roll" capabilities. Painted "FAA yellow," the apparatus is equipped with six-wheel drive and carries both FWD and Seagrave nameplates. Two tanks hold 2000 gallons of water and 75 gallons of foam. The front bumper features a remote controlled turret and undercarriage nozzles. An amber strobe sits atop the roof-mounted light bar. *Photo by Kent Parrish*

MODEL HP-97-DD, SERIAL NUMBER P-75422 – Pawcatuck, Connecticut received this 100-foot Rear Admiral with Detroit Diesel 8V-71TA engine in 1983. The tandem axle configuration allowed for a heavier payload. Three diamond plate cabinets adorned the rear of the driver's side with loose equipment stored in between. The apparatus was also equipped with a PTO-driven booster pump, small water tank, and hose reel. The aerial turntable was still based on the original S-Model specifications, thus the ladder rode with a slant on the full height cabs.
Photo by Dick Bartlett

MODEL MB-30-DA, SERIAL NUMBERS R-79470 THROUGH R-79481 – The M-Series pumpers were released to civilian departments in 1983 as the "Invader." The line retained its basic construction and option list to keep costs down. Philadelphia, Pennsylvania received the largest single municipal order with 12 units in 1984. Powered by Detroit Diesel 6-71N engines, they were equipped with 1000-gpm pumps and carried 500 gallons of water. The department received another three Invaders the following year. "Philly" was an excellent Seagrave customer, with no less than 90 deliveries during the Modern Era. *Photo by Jack Wright*

MODEL HD-50-DD, SERIAL NUMBER Q-79369 – Cedarburg, Wisconsin received the first Seagrave to have a raised roof cab and the only H-Model unit to see this option. The 1984 delivery featured four-wheel drive and carried a FWD nameplate. Powered by a Detroit Diesel 8V-71TA engine, it was equipped with a 1500-gpm pump and 750-gallon water tank. The pump area carried dual crosslays and a pre-piped deck gun with extension. Note the fire extinguishers and chrome trim band at the front. *Photo by Paul Barrett*

MODEL HD-44-DC, SERIAL NUMBER P-75424 – Seagrave produced a handful of these short-wheelbased Rear Admiral 75-foot quints. The standard aerial turntable was deleted and the aerial controls were placed at the pump panel. H-style outriggers and a single jack underneath the front bumper provided stabilization. This 1984 model took the shortest delivery trip possible, right through town to the Clintonville Volunteer Fire Department. Equipped with four-wheel drive and a Detroit Diesel 8V-71N engine, the rig also carried FWD plates. It had a 1250-gpm pump and 300-gallon water tank. The aerial controls were actually placed in a top-mount position. *Photo by Paul Barrett*

MODEL WR-54-DH, SERIAL NUMBER S-75479 – Painted chrome yellow with white upper cab, this 1985 delivery went to Federal Heights, Colorado. Powered by a Detroit Diesel 8V-92TA, it had a 1500-gpm pump and 300-gallon water tank. Aerial controls were placed in a little nook to the right of the master pump gauges. These jobs boasted pre-piped ladders with 500-pound tip loads. Ground ladders were stored in a bank behind the driver's side compartments and in standard fashion on the officer's side. *Photo by Dennis J. Maag*

69

MODEL HB-30-CF, SERIAL NUMBERS S-79564 THROUGH S-79579 – Los Angeles County received 16 of these 1000-gpm pumpers with 500-gallon water tanks in 1985 and another one the following year. They were powered by Cummins NTC-350 engines and had manual transmissions. Pictured is "LA Co." Engine 8, stationed in West Hollywood. The pump area demanded attention with its big Stang gun, crosslays, and dual booster reels. H-Model deliveries to California were consistent and similar in specifications. *Photo by Chuck Madderom*

MODEL HT-08-CF, SERIAL NUMBER S-76189 – This page highlights two different types of Seagrave 100-foot tractor-drawn aerials. The first is a 1985 delivery for Phoenix, Arizona with Cummins NTC-350 engine. The aerial has a pre-piped waterway and a 400-pound tip load. This configuration required a total of four swing-out stabilizer arms. The fully enclosed four-door cab is equipped with air conditioning, a comfort welcomed in the torrid desert heat. The trailer is fully compartmented. *Photo by Dave Greenberg*

MODEL HT-07-DH, SERIAL NUMBER S-76196 – The second example is a 1986 delivery with Detroit Diesel 8V-92TA engine for the St. Matthews Fire Protection District in Louisville, Kentucky. The aerial is a standard model with 200-pound tip load. This configuration only required two A-Frame stabilizers. The trailer was equipped with large compartments below the enclosed tiller cab. Note the deck gun mounted on the cab roof. *Photo by Kent Parrish*

MODEL HE-40-DH, SERIAL NUMBER T-79658 – The Chestnut Ridge Volunteer Fire Company in Maryland received this big tandem axle pumper-tanker in 1986. It had a Detroit Diesel 8V-92TA engine, 1250-gpm pump and 2500-gallon water tank. The rig sported a "Chicago-style" black-over-red livery. This scheme would become wildly popular across the country in the late 1990s. The pump panel was taller than normal and was equipped with an abundance of crosslays. The rear of the body had slanted beavertail compartments. *Photo by Howard Meile*

MODEL HB-51-DH, SERIAL NUMBER V-79773 – The McMahan Fire District in Louisville, Kentucky received this 1500-gpm pumper equipped with a 54-foot articulating Squrt in 1987. The fully enclosed cab was equipped with air-conditioning. Powered by a Detroit Diesel 8V-92TA engine, the unit carried 500 gallons of water on a short-wheelbase. McMahan's station is loaded with Seagraves. The current fleet also includes two 1994 J-Models, a 100-foot ladder and another Squrt; plus a 2004 T-Model 75-foot Meanstick quint. *Photo by Kent Parrish*

MODEL WP-59-DH, SERIAL NUMBER V-75533 – Seagrave introduced the RA110 heavy-duty aerial in 1987. This ladder was 110-feet in height and boasted a 550-pound tip load. It required a tandem axle and four H-style stabilizers. Northlake, Illinois received the first unit, a 1500-gpm quint with 350-gallon water tank and Detroit Diesel 8V-92TA engine. Note the front overhang of the aerial. The "cage" was to protect the pre-piped waterway from being damaged in rescue efforts and it was painted "day-glow" red so the operator could see the tip during low visibility situations. *Photo by Dennis J. Maag*

MODEL JB-50-DF, SERIAL NUMBER Y-78254 – In 1988, Seagrave introduced the J-Model or "Commander II." It featured a more modern, squared look than the H-Model and was widened to 96 inches. Changes included vertical side air intakes and improvements to the interior. Monroe-Lakeside, New York received this six-wheel drive pumper-tanker with FWD nameplates in 1988. It had a Detroit Diesel 6V-92TA engine, 1500-gpm pump, and 1500-gallon water tank. Booster reels were becoming ancient history; "bumper lines" would begin to handle the smaller fires. *Photo by Mike Martinelli*

MODEL HB-50-DF, SERIAL NUMBERS W-79810 THROUGH W-79818 – Houston, Texas was a good Seagrave customer during the 1980s with no less than 40 deliveries. The department received nine of these H-Model pumpers in 1988. They were equipped with Detroit Diesel 6V-92TA engines, 1500-gpm pumps, and 500-gallon water tanks. They had unusual compartmentation and carried combination ladders on the driver's side. *Photo by Eric Hansen*

MODEL WB-60-DH, SERIAL NUMBER W-79796 – The Rescue Fire Company in Cambridge, Maryland received this unusual W-Model pumper in 1988. Engine 3 had a Detroit Diesel 8V-92TA engine, big 1750-gpm pump, and 750-gallon water tank. The top-mount pump panel allowed the engineer to safely get off the street and have a 360-degree view of the fire scene. The body featured rescue-style compartments and pullout stands for upper access. *Photo by Howard Meile*

MODEL JT-07-CD, SERIAL NUMBER W-76237 – Honolulu, Hawaii was a good Seagrave customer. Ladder 29 at the Citron Street Station received this J-Model 100-foot tractor-drawn aerial in 1988. The open tiller seat didn't seem to be a bother in the gorgeous Pacific weather. Deliveries to Honolulu were powered by Cummins NTC-300 engines. They were painted "school-bus yellow" and were equipped with chrome wheels. Some Honolulu ladder companies even carried surfboards for ocean rescues! *Photo by Shaun P. Ryan*

MODEL JC-00-CD, SERIAL NUMBERS X-74194 & X-74195 – When Honolulu went shopping for two new rescue units, they wanted to stay with a Seagrave chassis for fleet standardization. Utilizing 1989 J-Model chassis, Marion Body Works completed these Search & Rescue rigs. The "walk-in" bodies allowed for a variety of functions. They could serve as command posts, rehab areas, and provide for additional storage. A towing package allowed the rigs to pull rescue boats on trailers. *Photo by Shaun P. Ryan*

MODEL JB-50-DH, SERIAL NUMBER X-79961 – Riverside, California received this 1989 delivery with top-mount pump panel. Engine 11 had a Detroit Diesel 8V-92TA engine, 1500-gpm pump, and 500-gallon water tank. This view displays another new apparatus trend. A ladder rack could be hydraulically lowered to an ergonomic level for ladder removal. This configuration allowed for more storage space, as engine companies were carrying more equipment than ever. Note the cutout at the pump panel for the booster reel. *Photo by Chuck Madderom*

MODEL LP-09-DH, SERIAL NUMBER Y-75613 – In 1990, Seagrave replaced the W-Model with the more modern L-Model. This cab still had the low profile forward section, but an offset roofline was added to the crew section to make for an ergonomic four-door model. The center of the upper cab featured a cutout for an aerial device to nest in. Unlike the W-Model, the L-Models have been exclusively utilized for aerial apparatus. Atlantic City, New Jersey received this 1990 RA110 with Detroit Diesel 8V-92TA engine. *Photo by Jack Wright*

MODEL JE-41-DH, SERIAL NUMBER X-79973 – Seagraves of this vintage equipped with Snorkels are rare. South Amboy, New Jersey received this 85-foot model in 1990 with Seagrave bodywork. Powered by a Detroit Diesel 8V-92TA engine, the apparatus carries a 1250-gpm pump, but no water tank. Note the air horns and electronic siren speaker recessed into the body. This would be the last year for canopy cabs. New NFPA standards would require all new apparatus to have fully enclosed cabs and seated positions for all firefighters. *Photo by Mike Martinelli*

MODEL JB-51-DF, SERIAL NUMBER Z-78345 – Seagrave delivered this 1991 pumper to the Buechel Fire District in Louisville, Kentucky. Powered by a Detroit Diesel 6V-92TA engine, the apparatus was equipped with a 1500-gpm pump and 500-gallon water tank. It also had a 54-foot articulating Squrt, which was a popular option on Louisville-area Seagraves. *Photo courtesy of Dennis Downes*

MODEL JB-40-DH, SERIAL NUMBER Y-78331 – The Deer Park-Silverton Joint Fire District in Ohio operates this 1991 heavy-duty rescue-pumper. Rescue 89 has a Detroit Diesel 8V-92TA engine and 1250-gpm pump. It carries 750 gallons of water and 100 gallons of foam. Both sides of the rescue body are equipped with "coffin-style" compartments that are accessed via hatches on top of the body. Ground ladders are stored internally. The new rescue-pumper trend gave smaller departments an extremely versatile first-out piece. *Photo by Greg Stapleton*

MODEL TB-50-DH, SERIAL NUMBER Y-78276 – In 1991, Seagrave unveiled its first custom tilt-cab. The T-Model or "Marauder" was available in full-tilt or split-tilt configurations. The split-tilt version had an extended cab with raised roof option and seating capacity of up to 10. The first delivery went to the Reliable Hose and Engine Company of Riverhead, New York. It had a Detroit Diesel 8V-92TA engine and 1500-gpm pump. In addition to a 750-gallon water tank, the unit carried 20-gallons each of Class A and B foam. *Photo by Mike Martinelli*

MODEL JP-02-DH, SERIAL NUMBER Z-75902 – In 1991, Seagrave introduced its first custom tower ladder. The 105-foot "Apollo" boasted an impressive tip load rating of 1000-pounds dry and 500-pounds while flowing water. The three-section aerial was constructed of high tensile steel and could be elevated from negative 5 to 80 degrees. The first delivery went to Prince George's County, Maryland. Painted white with a red stripe and equipped with dual monitors in the bucket, the apparatus had a Detroit Diesel 8V-92TA engine and was assigned to Glenn Dale Station 18. *Photo by Howard Meile*

MODEL JB-50-DH, SERIAL NUMBERS Z-78392 THROUGH Z-78394 – This pumper was one of three delivered to Bremerton, Washington in 1992. Each had a Detroit Diesel 8V-92TA engine, air-conditioned cab, 1500-gpm pump, and 500-gallon water tank. They were painted an attractive "school-bus yellow" with white upper cabs and red striping and lettering. Ladders were stored on a hydraulic rack. The rear fender plates were fashioned in diamond plate. *Photo by Bill Hattersley*

MODEL JP-06-DH, SERIAL NUMBER 75665 – New NFPA standards would require a minimum 250-pound tip load rating on all new aerial ladders. The venerable Seagrave Rear Admiral would be retired. The "Patriot," a new ladder with the required rating, was developed. It was available in 75- and 100-foot models, in both rear-mount and tractor-drawn configurations. Oak Harbor, Washington received this tandem axle 100-foot unit with Detroit Diesel 8V-92TA engine in 1992. A new style of warning light was becoming popular, the Federal Vector light pods. *Photo by Shane MacKichan*

MODEL TC-00-CH, SERIAL NUMBERS 74201 THROUGH 74203 – The Seagrave two-door tilt cab provided an excellent fit for specialty vehicles. Nearly a dozen have been built to date. Los Angeles, California received the first three such units. Saulsbury completed the bodies on these Hazardous-Materials Units in 1992. They were powered by Cummins NTC-400 engines. Extending over the front wheel wells, the walk-in bodies had office settings for Hazmat operations and crew seating, in addition to the large exterior storage compartments. *Photo by Chuck Madderom*

MODEL JB-30-CK, SERIAL NUMBERS 78453 THROUGH 78464 – This 1992 pumper was the pilot model for the new breed of FDNY 1000-gpm pumpers. The J-Model cab was slightly extended to allow firefighters in the forward facing flip seats to "pack up" while in route to a fire. Power was supplied via the Cummins L-10 engine. The unique body featured a "T" shaped 500-gallon water tank to allow for an extremely low hose bed. The officer's side was equipped with full depth upper compartments with roll-up doors and drop-down ladder racks. *Photo by John A. Calderone/Fire Apparatus Journal*

MODEL TB-30-CN, SERIAL NUMBER 78498 – New York City also operates seven high-pressure pumpers. They have 1000-gpm three-stage Waterous pumps, which are rated at 500-gpm at 700-psi when using the third stage. Engine 258 received this 1993 unit with Cummins N-14 engine and 500-gallon water tank. The FDNY would request this modified split-tilt cab on no less than 60 units over the course of several years; including pumpers, rear-mounts, tillers, and Aerialscopes. The department currently specifies the fixed-cab exclusively. *Photo by John A. Calderone/Fire Apparatus Journal*

MODEL TB-51-DH, SERIAL NUMBER 88008 – Woodstown, New Jersey operates this impressive 1993 Telesqurt-equipped pumper. It has a Detroit Diesel 8V-92TA engine, 1500-gpm pump, and 500-gallon water tank. An unusual feature for an aerial is the top-mount pump panel. Two "speedlays," a booster reel, and aerial controls round out the busy pump area. The black and white photo does not do this rig justice. The 65-foot boom is painted white with bright green tip. The apparatus has a lime scheme with dark green reflective striping and the side-mounted ground ladders are painted yellow. *Photo by Mike Martinelli*

MODEL LP-52-DH, SERIAL NUMBER 85001 – This 1993 105-foot Apollo went to North Brunswick, New Jersey. Powered by a Detroit Diesel 8V-92TA engine, it was equipped with a 1500-gpm pump and 150-gallon water tank. The L-Series low profile chassis was a nice fit for the Apollo. Note the air-conditioner condenser on the raised portion of the cab and the large power cord reel above the body. The Seagrave emblem between the driver and middle window had been updated—the flame was now embossed on a plate with black background. *Photo by Mike Martinelli*

MODEL JG-00-DH, SERIAL NUMBER 74205 – With Mack ceasing production of the CF-Series fire apparatus chassis, the popular Baker Aerialscope found itself without a partner. Seagrave would begin to supply its heavy-duty J-Model chassis under the FWD name. Saulsbury or Marion Body Works would complete the bodywork. Riverdale, Maryland received this 1993 model with Detroit Diesel 8V-92TA engine, 95-foot boom, and Saulsbury bodywork. It was painted white and black with gold striping. *Photo by Howard Meile*

MODEL TB-50-DH, SERIAL NUMBER 78654 – This green-over-white delivery went to the historic Independent Hose Company of Frederick, Maryland in 1994. The split-tilt extended cab featured a raised roof and could seat a crew of 10. Powered by a Detroit Diesel 8V-92TA engine, it was equipped with a 1500-gpm pump and 700-gallon water tank. Many new deliveries were featuring scene lights mounted on the side of the cab. Note the abundance of warning lights. *Photo by Howard Meile*

MODEL JP-56-DH, SERIAL NUMBERS 85007 & 85010 – Louisville, Kentucky is among Seagrave's best current customers. The department received two of these Patriot 100-foot quints in 1994. Powered by Detroit Diesel 8V-92TA engines, they were equipped with 1500-gpm pumps and 400-gallon water tanks. Note the lack of upper compartments on the officer's side. This was to accommodate additional storage of large diameter hose. Equipment is stored in every nook and cranny on these rigs. Louisville received a third quint two years later. *Photo by Kent Parrish*

MODEL TB-60-DA, SERIAL NUMBER 78714 – In 1995, the Eureka Fire Company in Stewartstown, Pennsylvania placed in service this 1750-gpm rescue-pumper with 1000-gallon water tank. It was powered by the new Detroit Diesel Series 60 engine. The apparatus was painted white with a wide orange band. The split-tilt extended cab featured a raised roof and could seat a crew of 10. The rescue body featured full height compartment doors and is equipped with coffin-style compartments. Ground ladders were stored internally. *Photo by Rick Rudisill*

MODEL LR-46-DH, SERIAL NUMBER 85015 – The Garrettford Drexel Hill Fire Company in Upper Darby Township, Pennsylvania operates this 1996 low profile 100-foot Patriot quint with pre-piped waterway. Powered by a Detroit Diesel 8V-92TA engine, the L-Model is equipped with a 1250-gpm pump and carries 300 gallons of water. The fire company also took delivery of a J-Model pumper at the same time. *Photo by Rick Rudisill*

MODEL TB-50-CN, SERIAL NUMBER 78738 – Los Angeles City, California received a total of 51 Seagrave tilt-cab pumpers between 1993 and 1996. Fifty were delivered with Cummins N-14 engines and a "test" rig was specified with a Detroit Diesel Series 60. All but two were equipped with manual transmissions, an uncommon option on fire apparatus of this vintage. Each unit had a 1500-gpm pump and 500-gallon water tank. One unit, painted lime, was equipped with a foam system and assigned to LAX International Airport. *Photo by Chuck Madderom*

MODEL JB-54-DH, SERIAL NUMBER 85025 – The Seagrave "P-75" was a pumper based model with a 75-foot rear-mount ladder. The standard aerial turntable was deleted and controls were located at the rear of the apparatus. Controls could also be placed at the pump panel. A standard complement of ground ladders was placed on the officer's side of the apparatus. The Camp Taylor Fire District in Louisville, Kentucky received this J-Model with Detroit Diesel 8V-92TA engine, 1500-gpm pump, and 500-gallon water tank in 1997. *Photo by Kent Parrish*

MODEL TT-06-DA, SERIAL NUMBER 76314 – After a 30-year absence, our Nation's Capital is once again protected by Seagrave. Washington, DC has purchased no less than 30 tilt-cab units since 1996. Truck 6 was assigned this 1997 100-foot tractor-drawn Patriot aerial with Detroit Diesel Series 40 engine. The department specified painted reinforced front bumpers on their apparatus. The first few units were painted white and red with stars, but the scheme has been changed to white-over-red. *Photo by Mike Martinelli*

MODEL JD-50-DH, SERIAL NUMBER 78788 – The Tahoe Douglas Fire District in Nevada protects a mountainous area with frequent snowfall. The department received this 1997 four-wheel drive unit, believed to be one of the last rigs so equipped by Seagrave, if not the last. Powered by a Detroit Diesel 8V-92TA engine, Engine 3 was equipped with a 1500-gpm pump and 750-gallon water tank. Remote controlled and heated bus-style mirrors were a convenience for the engineer. *Photo by Chuck Madderom*

MODEL TP-07-DA, SERIAL NUMBER 75732 – The Glyndon Volunteer Fire Department in Baltimore County, Maryland received a pair of 1997 Seagraves, a pumper and this well-appointed Patriot rear-mount. Although the ladder is lettered at 104 feet, it had a working height of 100 feet. The split-tilt extended cab could accommodate up to 10 firefighters. The powerplant was a Detroit Diesel Series 60 engine. Special undercarriage compartments were designed to carry air bags and cribbing. *Photo by Howard Meile*

MODEL JP-92-DA, SERIAL NUMBER 87006 – The Pleasure Ridge Park Fire Protection District in Louisville, Kentucky operates this 1997 Apollo 105-foot tower with Detroit Diesel Series 60 engine. Painted white-over-lime, the rig is equipped with a 250-gpm PTO-driven pump hidden in the front compartment and a 150-gallon water tank. A bumper line is tucked away in the front. This suburban department has purchased 13 rigs from Seagrave since 1991. *Photo by Kent Parrish*

MODEL JB-50-DH, SERIAL NUMBER 78860 – A blast from the past! Yes, this is a new quad. Louisville, Kentucky replaced an engine and truck with this rig in 1998. Quad 6 has a Detroit Diesel 8V-92TA engine, 1500-gpm pump, and 500-gallon water tank. Built on a long wheelbase, it has a large amount of compartment space to carry the necessary engine and truck company equipment. A full city complement of ground ladders is stored internally. The front grille provided more adequate cooling for the big engine. *Photo by Kent Parrish*

MODEL TR-64-DA, SERIAL NUMBER 85033 – The "Meanstick" has become one of Seagrave's most popular products. The 75-foot quint is designed to carry up to 500 gallons of water, a full complement of ground ladders, and 1000 feet of large diameter hose all on a short-wheelbase chassis with single rear axle. Ludlow, Kentucky received this 1998 delivery with Detroit Diesel Series 60 engine, 1750-gpm pump, and 400-gallon water tank. *Photo by David Mattingly*

MODEL JB-50-DF, SERIAL NUMBER 78907 – The Highview Fire District in Louisville, Kentucky received the last Seagrave with a Detroit Diesel 6V-92TA motor. New environmental regulations would promote the more compliant Detroit Diesel Series 40, 50, and 60 engines. This rescue-pumper had a 1500-gpm pump and 500-gallon water tank. The body had full height and depth compartments with five-bottle air cascade system and enclosed ladder storage. A hydraulic generator powered the 9000-watt Command Light mounted on the cab roof. *Photo by Kent Parrish*

MODEL JT-06-DA, SERIAL NUMBER 76319 – Paxtonia, Pennsylvania received a pair of Seagraves in 1998, a pumper and this 100-foot Patriot tractor-drawn aerial. Each was powered by a Detroit Diesel Series 60 engine and had the "short-door" option. Note the addition of roll-up compartment doors. The J-Model with front grille and eight-inch cab extension would become standard. It allowed for more interior legroom and a larger engine tunnel with improved cooling to accommodate the new engine models. *Photo by Greg Stapleton*

MODEL TB-70-DA, SERIAL NUMBER 78936 – Sycamore Township, Ohio was the recipient of this unique "patient transport pumper" in 1999. With a local hospital nearby, the department wanted the ability to quickly transport a patient or injured firefighter via their pumper if all of its ambulances were tied up. It was equipped with full height and depth compartments, Detroit Diesel Series 60 engine, 2000-gpm pump, 750-gallon water tank, and 30-gallon foam cell. The department had actually received a more primitive version three years earlier. *Photo by Kent Parrish*

A third cab door opened to allow for safe gurney operations. Cabinetry above the gurney was for storage of medical gear. The rear crew wall actually resembled the inside of an ambulance. *Photo by Kent Parrish*

107

MODEL JG-00-DA, SERIAL NUMBERS 74263 THROUGH 74267 – By now Seagrave had purchased the Aerialscope line from Baker. New York City received five 75-foot Aerialscopes in 1999 with Marion bodywork and Detroit Diesel Series 60 engines. FDNY Seagraves have many features unique to their specifications. Visible features include recessed handrails and exterior lights in rubber mounts. These options are meant to minimize the opportunity of field damage and lessen possible replacement costs. *Photo by John A. Calderone/Fire Apparatus Journal*

MODEL TB-80-DA, SERIAL NUMBER 78940 – The Humane Fire Company in Pottsville, Pennsylvania received this stout looking pumper in 1999. Powered by a Detroit Diesel Series 60 engine, this rig can pump a whopping 2250 gallons per minute. It carries 500 gallons of water and 30 gallons of foam. The apparatus is equipped with no less than six quartz scene lights and an FDNY-style hose bed. *Photo by Mike Martinelli*

MODEL TP-75-DA, SERIAL NUMBER 85040 – In 1999, Seagrave introduced the "Force" 100-foot aerial with 500-pound tip load. It was designed with four stabilizers; two narrow H-types at the rear and two modified A-frames behind the cab. The Fern Creek Fire Department in Louisville, Kentucky was the recipient of the pilot model. It is equipped with a Detroit Diesel Series 60 engine, 2000-gpm pump, and 300-gallon water tank. *Photo by Kent Parrish*

MODEL TB-51-DA, SERIAL NUMBER 78952 – Seagrave has finished at least three units equipped with 50-foot Snozzles. LaGrange, Kentucky operates this 2000 delivery with Detroit Diesel Series 60 engine, 1500-gpm pump, and 700-gallon water tank. It also carries 30 gallons of Class A and 20 gallons of Class B foam. A new option exclusive to Seagrave was a cab side access compartment, which occupied the dead space behind the two front seats. It was ideal for the storage of engineer's gear or medical equipment. *Photo by Kent Parrish*

MODEL TT-96-DA, SERIAL NUMBER 76321 – Seagrave constructed this impressive tractor-drawn 100-foot quint for the Empire Hook & Ladder Company of Upper Nyack, New York in 2000. It was Seagrave's first tillered aerial with a raised roof cab and the first with a bi-fold door in the tiller cab. The tractor was equipped with a Detroit Diesel Series 60 engine, 250-gpm PTO-driven pump, 200-gallon water tank, and dual crosslays. The trailer is loaded with wooden ladders, which are very uncommon for this part of the country. *Photo by Mike Martinelli*

MODEL TT-00-DA, SERIAL NUMBER 76334 – Orange, California was the recipient of this monstrous hybrid. Utilizing a 2000 Seagrave tractor and trailer with Detroit Diesel Series 60 engine, Super-Vac completed this tillered heavy rescue. Squad 6 is equipped with a Lima 40-kilowatt PTO generator, two 9000-watt Command Lights, and a Bauer air system. In addition to technical rescue equipment, over 123-feet of ground ladders are carried on drop-down racks. Note the full width tiller cab. *Photo by Chuck Madderom*

MODEL TP-54-DA, SERIAL NUMBER 85051 – Hadley, Massachusetts received this tandem axle 75-foot Patriot quint in 2000. Painted chrome yellow with white upper cab, the unit has a Detroit Diesel Series 60 engine, 1500-gpm pump, and 500-gallon water tank. Narrow rear outriggers stabilize the apparatus. Note the length of hard suction hose carried, quite uncommon for modern aerial apparatus. *Photo by Dick Bartlett*

MODEL JP-06-DA, SERIAL NUMBER 75776 – New York City lost over 30 pieces of Seagrave fire apparatus in the tragic events of September 11, 2001, and many more saw considerable damage. Seagrave and its employees went into overdrive to finish FDNY rigs that were near completion and to begin production on emergency orders. Units were completed in record time. Among the first deliveries was this 100-foot rear-mount with Detroit Diesel Series 60 engine for Ladder 10. Seagrave employees designed and donated the symbolic mural that adorned both sides of the body. Each replacement apparatus would feature a bronze medallion to honor the fallen firefighters.
Photo by John A. Calderone/Fire Apparatus Journal

MODEL JP-72-DA, SERIAL NUMBER 89012 – This 105-foot Apollo tower was delivered to Utica, New York in 2001. Powered by a Detroit Diesel Series 60, Tower 2 is equipped with a 2000-gpm pump and carries 300 gallons of water. It can mount quite an aerial assault with the dual monitors mounted on the platform. The apparatus is painted a mean looking black-over-red with black reflective striping. Note the mix of roll-up and standard compartment doors. *Photo by Mike Martinelli*

MODEL TB-00-DA, SERIAL NUMBER 78A02 – In 2001, Seagrave introduced the "Specialist," its own line of custom rescue vehicles. Models would be available with several cab options and body lengths of 14 to 22 feet. Construction consisted of Galvanneal or stainless steel. Compartment and shelving options were abundant, along with a long list of electrical and lighting packages. Sycamore Township, Ohio was again the recipient of a Seagrave first, with this 22-foot model. It was powered by Detroit Diesel Series 60 engine. *Photo by Kent Parrish*

MODEL TB-50-DC, SERIAL NUMBER 78A33 – Seagrave introduced the "Flame" in 1999. It was marketed as an all aluminum economy pumper targeted at communities that wanted an affordable custom rig. Slow to gain acceptance, very few orders were completed. The Levittown Fire Company in Pennsylvania received this 2001 model with Detroit Diesel Series 40 engine, 1500-gpm pump, 750-gallon water tank, and painted roll-up doors. *Photo by Dennis Sharpe*

MODEL TB-50-DC, SERIAL NUMBER 78A60 – The price of the Flame was minimized by simple construction and basic options. A first for Seagrave was a contoured wraparound front dash. Little Miami Fire & Rescue in Ohio purchased this 2002 demonstrator model with Detroit Diesel Series 40 engine, raised-roof cab, 1500-gpm pump, 750-gallon water tank, and hydraulic ladder rack. The body featured roll-up doors with the lower compartments extending completely to the rear. *Photo by Steve Hagy*

MODEL JG-00-CM, SERIAL NUMBER 74296 – Seagrave was now completing the bodywork on Aerialscopes in house. In 2003, production of the platform would be moved from Richmond, Virginia to the Clintonville plant. The Progress Fire Company in Susquehanna Township, Pennsylvania received this ornate 95-foot model in 2002 with Cummins M-11 engine. The upper half and lower quarter of the cab are painted red with the middle being white. A white reflective stripe runs through the red body. Note the classic placement of the Roto-Ray light. *Photo by Rick Rudisill*

MODEL TB-50-DA, SERIAL NUMBERS 78A44, 78A45, & 78A56 – Louisville, Kentucky received three of these Hazmat pumpers in 2002. Each had a Detroit Diesel Series 60 engine, 2000-gpm pump, and 500-gallon water tank. Dual foam systems could draw from 20-gallon Class A and 200-gallon Class B tanks. The standard Seagrave body was 96-inches wide. In order to store as much gear as possible, the bodies on these rigs were built 99-inches wide with a "flat back" design. Hydraulic racks stored ladders up and out of the way. *Photo by Kent Parrish*

MODEL LR-06-DA, SERIAL NUMBER 75780 – Baltimore, Maryland specified this low-profile 100-foot Patriot rear-mount for an older firehouse with space constrictions. It was delivered in 2002 and featured a Detroit Diesel Series 60 engine. Most big city departments still prefer traditional truck companies to quints. Many "truckies" jokingly refer to engine company crews as "speed bumps" as they muscle their way into a structure. *Photo by Howard Meile*

MODEL JB-40-DD, SERIAL NUMBER 78A58 – The Milford Community Fire Department in Ohio decided there was no better way to pay tribute to the FDNY than to purchase a "clone" pumper. The only basic differences on this 2003 delivery include a Detroit Diesel Series 50 engine, chrome front bumper, and upgraded 1250-gpm pump. This view shows the rear hose bed and officer's side, unique to FDNY specifications. *Photo by Kent Parrish*

MODEL JB-40-DD, SERIAL NUMBER 78A96 – Neighboring Mariemont, Ohio purchased its own clone the following year. They opted for the same powerplant and pump as the Milford delivery. A 30-gallon foam tank was added as well. Visible differences include the front Roto-Ray light and high side compartments on the driver's side. The two rigs on this page are the only ones Seagrave has built to these specifications for departments other than the FDNY. *Photo by Kent Parrish*

MODEL TR-75-DA, SERIAL NUMBER 85069 – In early 2003, Wallington, New Jersey placed in service one of the most striking rigs ever constructed by Seagrave. The entire upper half of the apparatus, including the pump panel and the 100-foot aerial ladder, is painted black. Truck 1 has a split-tilt extended cab that can accommodate a crew of 10. The Force aerial has a Detroit Diesel Series 60 engine, 2000-gpm pump, and 500-gallon water tank. *Photo by Mike Martinelli*

MODEL TE-50-CM, SERIAL NUMBER 78A82 – Somerset, New Jersey received this impressive Seagrave in 2003. The tandem axle pumper-tanker has a 1500-gpm pump, 1800-gallon water tank, 20-gallon foam cell, and Cummins M-11 engine. The white-over-lime apparatus has a full-tilt extended cab with raised roof and painted roll-up compartment doors. Hydraulic racks on both sides carry a folding dump tank, hard suction hoses, and various ladders. *Photo by Mike Martinelli*

MODEL TR-54-DA, SERIAL NUMBER 85073 – The Louisville Metro area in Kentucky has become one of Seagrave's most consistent sales territories. The Meade County Fire Protection District purchased this 2003 demonstrator model. It has a Detroit Diesel Series 60 engine, 1500-gpm pump, and 400-gallon water tank. It has full depth and height compartments on the driver's side with internal ladder storage on the officer's side. Note the unique flame graphics. *Photo by Kent Parrish*

125

MODEL JB-30-DA, SERIAL NUMBER 91768 – On the second anniversary of the September 11th tragedy, Seagrave and its employees donated this 1000-gpm rescue-pumper with Detroit Diesel Series 60 engine and 500-gallon water tank to Squad 61 of the FDNY. The body features a patriotic mural with the words "Never Forget," a tribute to the 343 firefighters that were killed. The formal event also unveiled the completely redesigned Commander II. Cab length was increased by six inches, upgrades were made in the engine compartment, and the chassis was improved. Visible changes include much larger windows and the disappearance of the side air intakes. This improved design is just a glimpse into Seagrave's aggressive plans for the future. *Photo by Mike Martinelli*

More Great Titles From Iconografix

All Iconografix books are available from direct mail specialty book dealers and bookstores worldwide, or can be ordered from the publisher. For book trade and distribution information or to add your name to our mailing list and receive a **FREE CATALOG** contact:

Iconografix, Inc.
PO Box 446, Dept BK
Hudson, WI, 54016

Telephone: (715) 381-9755,
(800) 289-3504 (USA),
Fax: (715) 381-9756

*This product is sold under license from Mack Trucks, Inc. Mack is a registered Trademark of Mack Trucks, Inc. All rights reserved.

EMERGENCY VEHICLES

The American Ambulance 1900-2002: An Illustrated History	ISBN 1-58388-081-X
American Fire Apparatus Co. 1922-1993 Photo Archive	ISBN 1-58388-131-X
American Funeral Vehicles 1883-2003 Illustrated History	ISBN 1-58388-104-2
American LaFrance 700 Series 1945-1952 Photo Archive	ISBN 1-882256-90-5
American LaFrance 700 Series 1945-1952 Photo Archive Volume 2	ISBN 1-58388-025-9
American LaFrance 700 & 800 Series 1953-1958 Photo Archive	ISBN 1-882256-91-3
American LaFrance 900 Series 1958-1964 Photo Archive	ISBN 1-58388-002-X
Classic Seagrave 1935-1951 Photo Archive	ISBN 1-58388-034-8
Crown Firecoach 1951-1985 Photo Archive	ISBN 1-58388-047-X
Encyclopedia of Canadian Fire Apparatus	ISBN 1-58388-119-0
Fire Chief Cars 1900-1997 Photo Album	ISBN 1-882256-87-5
Hahn Fire Apparatus 1923-1990 Photo Archive	ISBN 1-58388-077-1
Heavy Rescue Trucks 1931-2000 Photo Gallery	ISBN 1-58388-045-3
Imperial Fire Apparatus 1969-1976 Photo Archive	ISBN 1-58388-091-7
Industrial and Private Fire Apparatus 1925-2001 Photo Archive	ISBN 1-58388-049-6
Mack Model C Fire Trucks 1957-1967 Photo Archive*	ISBN 1-58388-014-3
Mack Model L Fire Trucks 1940-1954 Photo Archive*	ISBN 1-882256-86-7
Maxim Fire Apparatus 1914-1989 Photo Archive	ISBN 1-58388-050-X
Maxim Fire Apparatus Photo History	ISBN 1-58388-111-5
Navy & Marine Corps Fire Apparatus 1836 -2000 Photo Gallery	ISBN 1-58388-031-3
Pierre Thibault Ltd. Fire Apparatus 1918-1990 Photo Archive	ISBN 1-58388-074-7
Pirsch Fire Apparatus 1890-1991 Photo Archive	ISBN 1-58388-082-8
Police Cars: Restoring, Collecting & Showing America's Finest Sedans	ISBN 1-58388-046-1
Saulsbury Fire Rescue Apparatus 1956-2003 Photo Archive	ISBN 1-58388-106-9
Seagrave 70th Anniversary Series Photo Archive	ISBN 1-58388-001-1
Seagrave Fire Apparatus 1959-2004 Photo Archive	ISBN 1-58388-132-8
TASC Fire Apparatus 1946-1985 Photo Archive	ISBN 1-58388-065-8
Volunteer & Rural Fire Apparatus Photo Gallery	ISBN 1-58388-005-4
W.S. Darley & Co. Fire Apparatus 1908-2000 Photo Archive	ISBN 1-58388-061-5
Wildland Fire Apparatus 1940-2001 Photo Gallery	ISBN 1-58388-056-9
Young Fire Equipment 1932-1991 Photo Archive	ISBN 1-58388-015-1

TRUCKS

Autocar Trucks 1899-1950 Photo Archive	ISBN 1-58388-115-8
Autocar Trucks 1950-1987 Photo Archive	ISBN 1-58388-072-0
Beverage Trucks 1910-1975 Photo Archive	ISBN 1-882256-60-3
Brockway Trucks 1948-1961 Photo Archive*	ISBN 1-882256-55-7
Chevrolet El Camino Photo History Incl. GMC Sprint & Caballero	ISBN 1-58388-044-5
Circus and Carnival Trucks 1923-2000 Photo Archive	ISBN 1-58388-048-8
Dodge B-Series Trucks Restorer's & Collector's Reference Guide and History	ISBN 1-58388-087-9
Dodge Pickups 1939-1978 Photo Album	ISBN 1-882256-82-4
Dodge Power Wagons 1940-1980 Photo Archive	ISBN 1-882256-89-1
Dodge Power Wagon Photo History	ISBN 1-58388-019-4
Dodge Ram Trucks 1994-2001 Photo History	ISBN 1-58388-051-8
Dodge Trucks 1929-1947 Photo Archive	ISBN 1-882256-36-0
Dodge Trucks 1948-1960 Photo Archive	ISBN 1-882256-37-9
Ford 4x4s 1935-1990 Photo History	ISBN 1-58388-079-8
Ford Heavy-Duty Trucks 1948-1998 Photo History	ISBN 1-58388-043-7
Ford Ranchero 1957-1979 Photo History	ISBN 1-58388-126-3
Freightliner Trucks 1937-1981 Photo Archive	ISBN 1-58388-090-9
GMC Heavy-Duty Trucks 1927-1987	ISBN 1-58388-125-5
Jeep 1941-2000 Photo Archive	ISBN 1-58388-021-6
Jeep Prototypes & Concept Vehicles Photo Archive	ISBN 1-58388-033-X
Mack Model AB Photo Archive*	ISBN 1-882256-18-2
Mack AP Super-Duty Trucks 1926-1938 Photo Archive*	ISBN 1-882256-54-9
Mack Model B 1953-1966 Volume 2 Photo Archive*	ISBN 1-882256-34-4
Mack EB-EC-ED-EE-EF-EG-DE 1936-1951 Photo Archive*	ISBN 1-882256-29-8
Mack EH-EJ-EM-EQ-ER-ES 1936-1950 Photo Archive*	ISBN 1-882256-39-5
Mack FC-FCSW-NW 1936-1947 Photo Archive*	ISBN 1-882256-28-X
Mack FG-FH-FJ-FK-FN-FP-FT-FW 1937-1950 Photo Archive*	ISBN 1-882256-35-2
Mack LF-LH-LJ-LM-LT 1940-1956 Photo Archive*	ISBN 1-882256-38-7
Mack Trucks Photo Gallery*	ISBN 1-882256-88-3
New Car Carriers 1910-1998 Photo Album	ISBN 1-882256-98-0
Plymouth Commercial Vehicles Photo Archive	ISBN 1-58388-004-6
Refuse Trucks Photo Archive	ISBN 1-58388-042-9
Studebaker Trucks 1927-1940 Photo Archive	ISBN 1-882256-40-9
White Trucks 1900-1937 Photo Archive	ISBN 1-882256-80-8

BUSES

Buses of ACF Photo Archive Including ACF-Brill And CCF-Brill	ISBN 1-58388-101-8
Buses of Motor Coach Industries 1932-2000 Photo Archive	ISBN 1-58388-039-9
Fageol & Twin Coach Buses 1922-1956 Photo Archive	ISBN 1-58388-075-5
Flxible Intercity Buses 1924-1970 Photo Archive	ISBN 1-58388-108-5
Flxible Transit Buses 1953-1995 Photo Archive	ISBN 1-58388-053-4
GM Intercity Coaches 1944-1980 Photo Archive	ISBN 1-58388-099-2
Greyhound in Postcards: Buses, Depots and Posthouses	ISBN 1-58388-130-1
Highway Buses of the 20th Century Photo Gallery	ISBN 1-58388-121-2
Mack® Buses 1900-1960 Photo Archive*	ISBN 1-58388-020-8
Prevost Buses 1924-2002 Photo Archive	ISBN 1-58388-083-6
Trailways Buses 1936-2001 Photo Archive	ISBN 1-58388-029-1
Trolley Buses 1913-2001 Photo Archive	ISBN 1-58388-057-7
Yellow Coach Buses 1923-1943 Photo Archive	ISBN 1-58388-054-2

AUTOMOTIVE

AMC Cars 1954-1987: An Illustrated History	ISBN 1-58388-112-3
AMC Performance Cars 1951-1983 Photo Archive	ISBN 1-58388-127-1
AMX Photo Archive: From Concept to Reality	ISBN 1-58388-062-3
Auburn Automobiles 1900-1936 Photo Archive	ISBN 1-58388-093-3
Camaro 1967-2000 Photo Archive	ISBN 1-58388-032-1
Checker Cab Co. Photo History	ISBN 1-58388-100-X
Chevrolet Corvair Photo History	ISBN 1-58388-118-2
Chevrolet Station Wagons 1946-1966 Photo Archive	ISBN 1-58388-069-0
Classic American Limousines 1955-2000 Photo Archive	ISBN 1-58388-041-0
Cord Automobiles L-29 & 810/812 Photo Archive	ISBN 1-58388-102-6
Corvair by Chevrolet Experimental & Production Cars 1957-1969, Ludvigsen Library Series	ISBN 1-58388-058-5
Corvette The Exotic Experimental Cars, Ludvigsen Library Series	ISBN 1-58388-017-8
Corvette Prototypes & Show Cars Photo Album	ISBN 1-882256-77-8
Early Ford V-8s 1932-1942 Photo Album	ISBN 1-882256-97-2
Ferrari- The Factory Maranello's Secrets 1950-1975, Ludvigsen Library Series	ISBN 1-58388-085-2
Ford Postwar Flatheads 1946-1953 Photo Archive	ISBN 1-58388-080-1
Ford Station Wagons 1929-1991 Photo History	ISBN 1-58388-103-4
Hudson Automobiles 1934-1957 Photo Archive	ISBN 1-58388-110-7
Imperial 1955-1963 Photo Archive	ISBN 1-882256-22-0
Imperial 1964-1968 Photo Archive	ISBN 1-882256-23-9
Javelin Photo Archive: From Concept to Reality	ISBN 1-58388-071-2
Lincoln Motor Cars 1920-1942 Photo Archive	ISBN 1-58388-57-3
Lincoln Motor Cars 1946-1960 Photo Archive	ISBN 1-882256-58-1
Nash 1936-1957 Photo Archive	ISBN 1-58388-086-0
Packard Motor Cars 1935-1942 Photo Archive	ISBN 1-882256-44-1
Packard Motor Cars 1946-1958 Photo Archive	ISBN 1-882256-45-X
Pontiac Dream Cars, Show Cars & Prototypes 1928-1998 Photo Album	ISBN 1-882256-93-X
Pontiac Firebird Trans-Am 1969-1999 Photo Album	ISBN 1-882256-95-6
Pontiac Firebird 1967-2000 Photo History	ISBN 1-58388-028-3
Rambler 1950-1969 Photo Archive	ISBN 1-58388-078-X
Stretch Limousines 1928-2001 Photo Archive	ISBN 1-58388-070-4
Studebaker 1933-1942 Photo Archive	ISBN 1-882256-24-7
Studebaker Hawk 1956-1964 Photo Archive	ISBN 1-58388-094-1
Studebaker Lark 1959-1966 Photo Archive	ISBN 1-58388-107-7
Ultimate Corvette Trivia Challenge	ISBN 1-58388-035-6

RECREATIONAL VEHICLES

RVs & Campers 1900-2000: An Illustrated History	ISBN 1-58388-064-X
Ski-Doo Racing Sleds 1960-2003 Photo Archive	ISBN 1-58388-105-0

RAILWAYS

Burlington Zephyrs Photo Archive: America's Distinctive Trains	ISBN 1-58388-124-7
Chicago, St. Paul, Minneapolis & Omaha Railway 1880-1940 Photo Archive	ISBN 1-882256-67-0
Chicago & North Western Railway 1975-1995 Photo Archive	ISBN 1-882256-76-X
Classic Streamliners Photo Archive: The Trains and the Designers	ISBN 1-58388-144-x
Great Northern Railway 1945-1970 Volume 2 Photo Archive	ISBN 1-882256-79-4
Great Northern Railway Ore Docks of Lake Superior Photo Archive	ISBN 1-58388-073-9
Illinois Central Railroad 1854-1960 Photo Archive	ISBN 1-58388-063-1
Locomotives of the Upper Midwest Photo Archive: Diesel Power in the 1960s and 1970s	ISBN 1-58388-113-1
Milwaukee Road 1850-1960 Photo Archive	ISBN 1-882256-61-1
Milwaukee Road Depots 1856-1954 Photo Archive	ISBN 1-58388-040-2
Show Trains of the 20th Century	ISBN 1-58388-030-5
Soo Line 1975-1992 Photo Archive	ISBN 1-882256-68-9
Steam Locomotives of the B&O Railroad Photo Archive	ISBN 1-58388-095-X
Streamliners to the Twin Cities Photo Archive 400, Twin Zephyrs & Hiawatha Trains	ISBN 1-58388-096-8
Trains of the Twin Ports Photo Archive, Duluth-Superior in the 1950s	ISBN 1-58388-003-8
Trains of the Circus 1872-1956	ISBN 1-58388-024-0
Trains of the Upper Midwest Photo Archive Steam & Diesel in the 1950s & 1960s	ISBN 1-58388-036-4
Wisconsin Central Limited 1987-1996 Photo Archive	ISBN 1-882256-75-1
Wisconsin Central Railway 1871-1909 Photo Archive	ISBN 1-882256-78-6

More great books from Iconografix

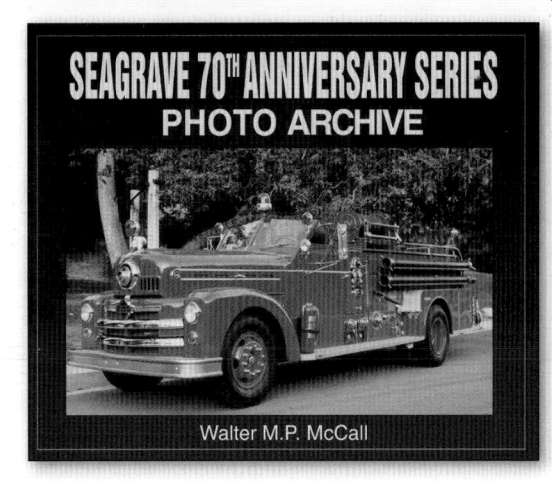

CLASSIC SEAGRAVE 1935 - 1951 PHOTO ARCHIVE
Walt McCall & Matt Lee
ISBN 1-58388-034-8

SEAGRAVE 70TH ANNIVERSARY SERIES PHOTO ARCHIVE
Walter M.P. McCall
ISBN 1-58388-001-1

AMERICAN FIRE APPARATUS CO. 1922-1993 PHOTO ARCHIVE
Richard J. Gergel
ISBN 1-58388-131-X

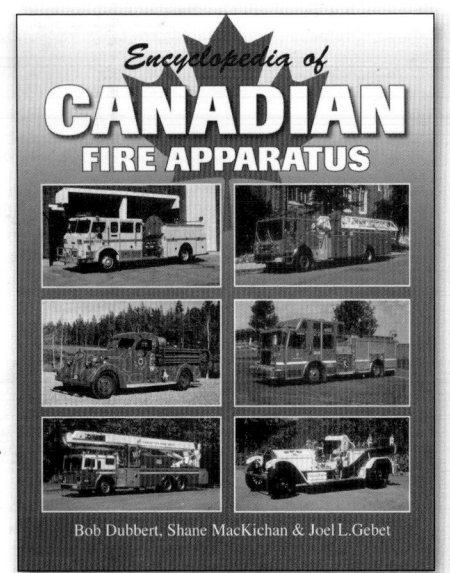

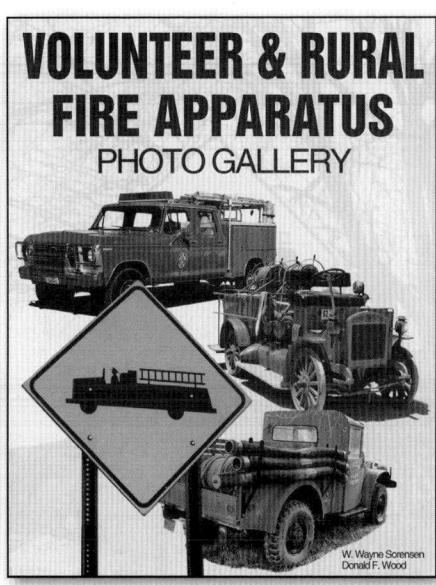

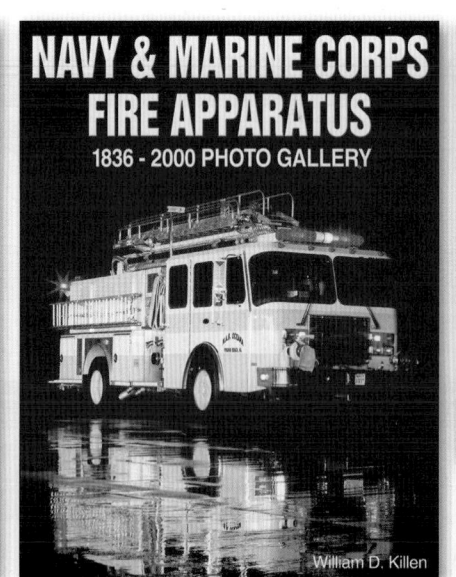

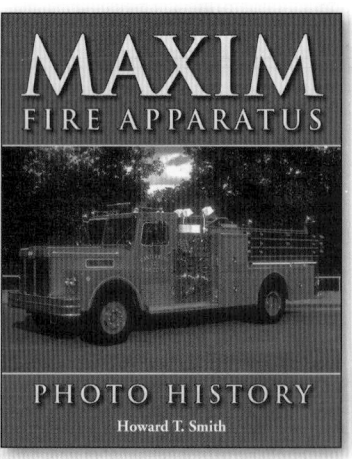

Encyclopedia of CANADIAN FIRE APPARATUS
Bob Dubbert, Shane MacKichan & Joel L. Gebet
ISBN 1-58388-119-0

VOLUNTEER & RURAL FIRE APPARATUS PHOTO GALLERY
W. Wayne Sorensen, Donald F. Wood
ISBN 1-58388-005-4

NAVY & MARINE CORPS FIRE APPARATUS 1836 - 2000 PHOTO GALLERY
William D. Killen
ISBN 1-58388-031-3

MAXIM FIRE APPARATUS PHOTO HISTORY
Howard T. Smith
ISBN 1-58388-111-5

Iconografix, Inc. P.O. Box 446, Dept BK, Hudson, WI 54016
For a free catalog call: 1-800-289-3504